Where on Earth?

Plains

by JoAnn Early Macken

Reading consultant: Susan Nations, M.Ed.,
author, literacy coach,
and consultant in literacy development

WEEKLY READER®
PUBLISHING

Learning from Maps

You can learn many things from maps if you know how to read them. This page will help you understand how to read a map.

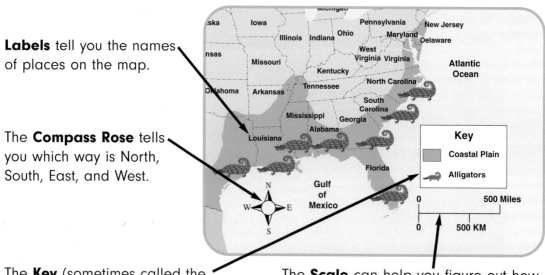

Labels tell you the names of places on the map.

The **Compass Rose** tells you which way is North, South, East, and West.

The **Key** (sometimes called the Legend) tells you what the symbols or special colors on the map mean.

The **Scale** can help you figure out how big or far apart places on the map are. For example, a distance of 1 inch (2.5 centimeters) on a map may be 100 miles (161 kilometers) in the real world.

Table of Contents

What Is a Plain? 4

Coastal Plains 6

Grasslands 8

Life on the Plains 16

People and Plains 20

Spotlight: The Serengeti 21

Glossary 22

For More Information......................23

Index 24

Cover and title page: Some plains are covered with grasses and wild flowers.

This plain has small hills and a few trees.

What Is a Plain?

A plain is a large area of land that is mostly flat. Some plains have small hills. Trees grow on some plains. Others are covered with grass.

Plains are found all over the world. They can be wet or dry. They can be cold or warm.

Glaciers helped shape some plains. A **glacier** is a river of ice. The weight of a glacier flattens the land as it moves.

Other plains formed when volcanoes **erupted**, or blew open. Lava flows filled in the low spots on these plains.

Some plains were once at the bottom of a sea. The water that covered them helped shape the land.

Grass covers this plain, but no trees grow here.

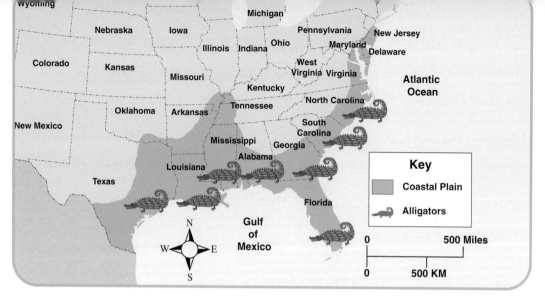

This map shows where American alligators are located on the coastal plain.

Coastal Plains

A plain that forms next to an ocean, a sea, or another large body of water is called a coastal plain. Rivers drop sand and soil where they reach the ocean. A coastal plain can form there.

Over time, the sea level rises and falls. It covers land, and then it leaves the land dry. Waves wash in and out. Water keeps changing the land.

Some coastal plains have wetlands and beaches. Many kinds of plants grow in these places. Trees grow on some coastal plains.

Water birds and turtles live on coastal plains. Alligators are found in some areas. Deer, bears, and foxes all find homes on coastal plains.

Water birds search for food. Coastal plains have many kinds of plants and animals.

This plain has just a few trees and bushes.

Grasslands

An **inland plain** is far from the ocean. Many inland plains are covered with grass. They are called grasslands. They are in places that are too dry for many trees to grow. Some are **tropical** grasslands. They are in hot places. Others are **temperate** grasslands. They are in places that do not get very hot or very cold.

All grasslands are alike in certain ways. Little rain falls on them. Rain may fall only at certain times of the year. Grasslands are windy because they have few trees to block the wind. From time to time, there may be droughts. **Droughts** are times when very little or no rain falls. They can cause problems for plants and animals.

Farmers have tried to grow crops on this plain in Nebraska. But a drought has kept the crops from growing.

Mountains rise up behind this plain. Although they have some snow on top, the plain does not get much water. Only grass grows there.

North America has huge plains in its center. They stretch eastward from the Rocky Mountains. These mountains are in the west. Damp winds reach them on the western side. These winds leave most of their water on the mountains. On the eastern side, the land is mostly dry.

The Mississippi River flows south through the middle of the plains. The land between the river and the Rockies is called the Great Plains. Most of it is covered by grassy prairies. The land closest to the Mississippi River has more water and long grasses. The land close to the Rockies is drier and has short grass. In between these two areas, the prairie has both short and long grasses.

In some grasslands, you can see beautiful wild flowers.

Grasslands can have different names. In Europe and Asia, they are called steppes. Grassy plains in Australia and New Zealand are called downs. In South America, they are sometimes called pampas.

Most grasslands in Africa are called savannas. Some are called velds. Savannas are tropical grasslands. They are found in hot places. Savannas have a dry season and a rainy season. They get enough rain for some trees and bushes to grow.

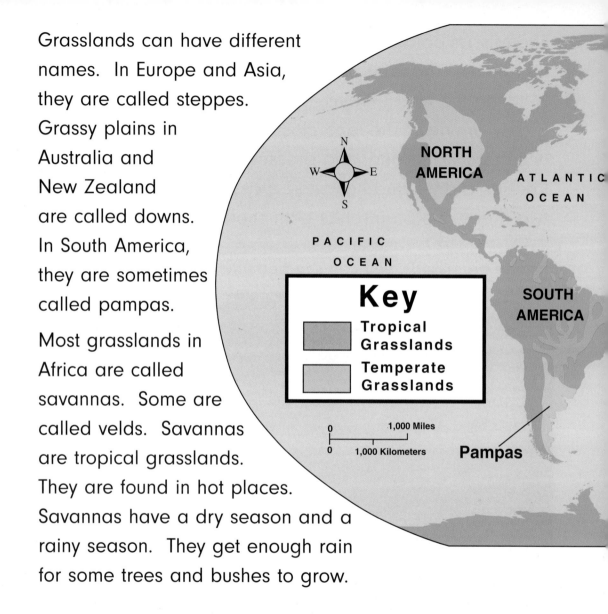

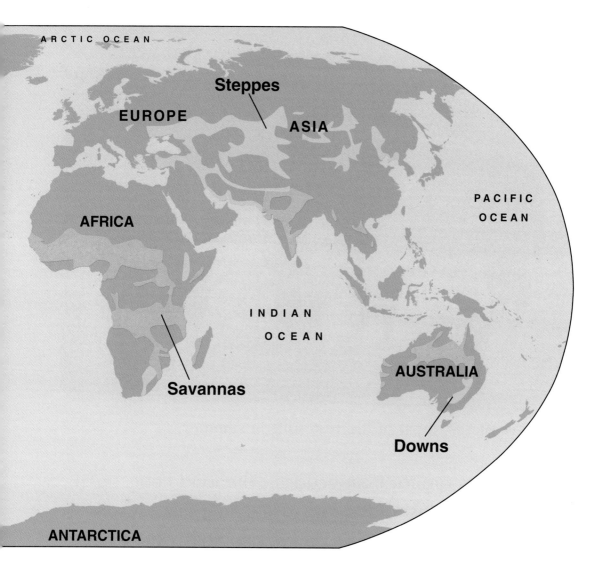

This map shows different grasslands in the world.

On some plains, most of the rain falls in summer.

Many plains are far from oceans. The land heats up in the summer. Most of the rain falls during this season. In winter, the weather can be freezing cold. Snow may fall in winter.

Grasslands burn from time to time. Lightning starts fires, which can spread easily. Grass burns above the ground, but the roots of the grass are not hurt. Ashes from burned plants make the soil richer. New grass grows soon after a fire.

When lightning strikes grasslands, dry grass can catch fire.

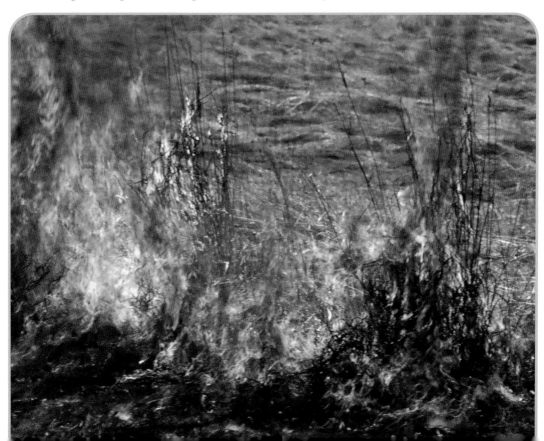

On grassy plains, the grass provides food for animals.

Life on the Plains

On some plains, rain does not fall often. But some grasses can still grow. Their deep roots reach water far under the ground. After animals eat the tops of the grass, it keeps growing. Wind blows their seeds around, and new grass grows from these seeds.

Small animals dig under the grasslands. On prairies, prairie dogs live underground. They live together in a "town." Thousands of prairie dogs may live in one town, which may be miles wide. They take turns watching for enemies. Small rodents also live in **burrows**, or holes, under steppes and pampas.

Prairie dogs guard their "town" from enemies.

Grazing animals feast on grasslands in Australia.

Large animals eat grass on the grasslands. Prairies are home to bison herds. On the plains of Africa, elephants and rhinos eat grasses. Kangaroos **graze** on, or eat, the grasslands of Australia.

Different birds can be found in grasslands. Ostriches live in Africa. Emus live in Australia. Rheas live in South America. In some ways, these birds are all alike. They are all large birds with long legs and small wings. They do not fly.

A mother ostrich watches her young.

On some plains, people work on large farms.

People and Plains

Many plains are now used for farming. Where glaciers formed the land, rich soil was left behind. Farmers on the prairie grow wheat and corn. On steppes, farmers grow oats and rye.

Some areas are too dry for farming. Cattle and sheep graze on the grass there.

Spotlight: The Serengeti

The Serengeti is an African savanna. Its name means "endless plain." It is located on the eastern side of Africa. In this savanna, zebras feed on grasses and leaves. Cheetahs hunt gazelles and other prey. They are the fastest land animals in the world. Wildebeests migrate across the savanna. They follow the rain, which brings more green grass.

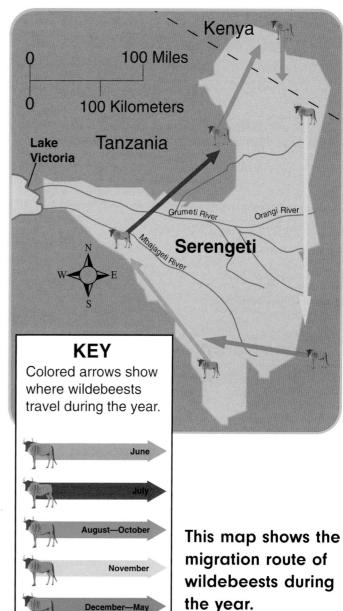

This map shows the migration route of wildebeests during the year.

Glossary

droughts — times when the land is very dry

erupted — burst or broken open in a very forceful way

glaciers — rivers of ice

lava — rock that is melted by the heat of Earth and then flows from a volcano. Lava hardens after it cools.

migrate — to make a journey from one place or climate to another

temperate — having to do with places on Earth that do not get very hot or very cold.

tropical — having to do with places on Earth that are hot year-round.

volcanoes — holes in Earth's crust where lava, rocks, and ashes flow from inside Earth

wetlands — places that stay wet most of the year, such as bogs, marshes, and swamps

For More Information

Books

Grasslands. Biomes of North America (series). Lynn M. Stone (Rourke)

Grassy Lands. We Can Read About Nature! (series). Catherine Nichols (Benchmark Books)

Living in the Savannah. Linda Bullock (Children's Press)

Living on a Plain. Communities (series). Joanne Winne (Children's Press)

Living on the Plains. Allan Fowler (Children's Press)

Prairie Dogs. Grassland Animals (series). Patricia J. Murphy (Capstone)

Web Sites

Grasslands Biome
mbgnet.mobot.org/sets/grasslnd
Learn about grasslands and the animals and plants that live there.

Plains Zebra
www.pbs.org/kratts/world/africa/zebra/index.html
Zebra facts from Kratt's Creatures

Index

animals 7, 9, 16, 17, 18, 19, 21
coastal plains 6, 7
droughts 9
glaciers 5, 20
grass 3, 4, 5, 8, 10, 11, 12, 13, 15, 16, 18, 20, 21
grasslands 8, 9, 11, 12, 13, 15, 17, 18, 19
inland plains 8
mountains 10, 11
people 20
plants 7, 9, 15
rain 9, 12, 14, 16, 21
rivers 6, 11
Serengeti 21
trees 4, 5, 7, 8, 9, 12
volcanoes 5
water 5, 6, 7, 10, 11, 16
winds 9, 10, 16

About the Author

JoAnn Early Macken is the author of two rhyming picture books, *Sing-Along Song* and *Cats on Judy*, and more than eighty nonfiction books for children. Her poems have appeared in several children's magazines. A graduate of the M.F.A. in Writing for Children and Young Adults Program at Vermont College, she lives in Wisconsin with her husband and their two sons.

Please visit our web site at: www.garethstevens.com.
For a free color catalog describing our list of quality books, call 1-800-542-2595 (USA) or 1-800-387-3178 (Canada).
Our fax: 1-877-542-2596

Library of Congress Cataloging-in-Publication Data

Macken, JoAnn Early, 1953–
  Plains / JoAnn Early Macken.
    p. cm. — (Where on earth? world geography)
  Includes bibliographical references and index.
  ISBN-10: 0-8368-6396-8   ISBN-13: 978-0-8368-6396-3 (lib. bdg.)
  ISBN-10: 0-8368-6403-4   ISBN-13: 978-0-8368-6403-8 (softcover)
  1. Plains—Juvenile literature. I. Title.
GB572.M33 2006
551.45'3—dc22   2005025520

This edition first published in 2006 by
Weekly Reader® Books
An Imprint of Gareth Stevens Publishing
1 Reader's Digest Road
Pleasantville, NY 10570-7000 USA

Copyright © 2006 by Weekly Reader® Early Learning Library

Editors: Jim Mezzanotte and Barbara Kiely Miller
Art direction: Tammy West
Cover design and page layout: Kami Strunsee
Picture research: Diane Laska-Swanke

Picture credits: Cover, title, © Ron Spomer/Visuals Unlimited; pp. 2, 6, 12-13, 21 Kami Strunsee/© Weekly Reader Early Learning Library, 2006; pp. 4, 5, 11, 15, 17, 18 © James P. Rowan; pp. 7, 10, 14 © Tom and Pat Leeson; p. 8 © Daniel Gomez/naturepl.com; pp. 9, 20 © Inga Spence/Visuals Unlimited; p. 16 © Jeremy Walker/naturepl.com; p. 19 © Joe McDonald/Visuals Unlimited

All rights reserved. No part of this book may be reproduced, stored in a retrieval system, or transmitted in any form or by any means, electronic, mechanical, photocopying, recording, or otherwise, without the prior written permission of the copyright holder.

Printed in the United States of America

2 3 4 5 6 7 8 9 10 09 08